Slip, Trip, and Fall Prevention for Healthcare Workers

DEPARTMENT OF HEALTH AND HUMAN SERVICES
Centers for Disease Control and Prevention
National Institute for Occupational Safety and Health

Authors

Jennifer Bell, Ph.D.
Research Epidemiologist
Division of Safety Research
National Institute for Occupational
   Safety and Health
Centers for Disease Control and Prevention
Morgantown, WV

James W. Collins, Ph.D., MSME
Associate Director for Science
Division of Safety Research
National Institute for Occupational
   Safety and Health
Centers for Disease Control and Prevention
Morgantown, WV

Elizabeth Dalsey, MA
Health Communication Specialist
Office of Research and Technology Transfer
National Institute for Occupational
   Safety and Health
Centers for Disease Control and Prevention
Cincinnati, OH

Virginia Sublet, Ph.D.
Senior Health Scientist
Office of Health Communication and Global
   Collaborations
National Institute for Occupational
   Safety and Health
Centers for Disease Control and Prevention
Washington, DC

This document is in the public domain and may be freely copied or reprinted.

Disclaimer

Mention of any company or product does not constitute endorsement by the National Institute for Occupational Safety and Health (NIOSH). In addition, citations to Web sites external to NIOSH do not constitute NIOSH endorsement of the sponsoring organizations or their programs or products. Furthermore, NIOSH is not responsible for the content of these Web sites.

Ordering Information

To receive documents or other information about occupational safety and health topics, contact NIOSH at

Telephone: 1–800–CDC–INFO (1–800–232–4636)
TTY: 1–888–232–6348
E-mail: cdcinfo@cdc.gov

or visit the NIOSH Web site at www.cdc.gov/niosh.

For a monthly update on news at NIOSH, subscribe to *NIOSH eNews* at www.cdc.gov/niosh/eNews.

DHHS (NIOSH) Publication Number 2011–123

December 2010

SAFER • HEALTHIER • PEOPLE™

Contents

Part I. Introduction 1

   Slip, Trip, and Fall (STF) Prevention for Healthcare Workers   3

Part II. Top 10 Hazards 7

   1. Contaminants on the Floor 9
   2. Poor Drainage: Pipes and Drains 14
   3. Indoor Walking Surface Irregularities 16
   4. Outdoor Walking Surface Irregularities 17
   5. Weather Conditions: Ice and Snow 19
   6. Inadequate Lighting 20
   7. Stairs and Handrails 21
   8. Stepstools and Ladders 23
   9. Tripping Hazards: Clutter, Loose Cords, Hoses, Wires, and Medical Tubing 24
   10. Improper Use of Floor Mats and Runners 27

Part III. Tools for STF Prevention 29

   Examine Employee Slips, Trips, and Falls Injuries: The First Steps Toward Prevention 31

   Employee Communication: Training and Involvement 34

   Slips, Trips, and Falls Checklist 35

   Literature Cited 40

Contributors and Reviewers

This work emerged from a multi-institutional, multi-disciplinary research partnership including NIOSH, BJC Health System, the Finnish Institute for Occupational Health, Liberty Mutual Research Institute for Safety, the Johns Hopkins School of Public Health, the Veteran's Health Administration, and Washington University School of Medicine.

The authors of this document would like to convey special thanks to the following individuals for their review of and contribution to the document:

Barbara I. Braun, Ph.D.
Project Director, Department of Health Services Research
Department of Health Services Research
The Joint Commission
Oakbrook Terrace IL

Theodore K. Courtney, MS, CSP
Director, Center for Injury Epidemiology
Liberty Mutual Research Institute for Safety
Hopkinton, Massachusetts

Kathy Gerwig
Vice President, Workplace Safety and Environmental Stewardship
Kaiser Permanente
Oakland, CA

Joshua M. Harney, MS, CIH
Director, Occupational Safety & Environmental Health
Cincinnati Children's Hospital Medical Center
Cincinnati, OH

Michael J. Hodgson, MD, MPH
Chief Consultant, Occupational Health,
Veteran's Health Administration
Washington, DC

Wayne S. Maynard, CSP, CPE, ALCM
Manager, Technical Services &
  Product Development
Ergonomics and Tribology
Liberty Mutual Group
Hopkinton, MA

Gary Sorock, Ph.D.
Adjunct Associate Professor
Center for Injury Research and Policy
Johns Hopkins Bloomberg School
  of Public Health
Glyndon, MD

Laurie Wolf, MS, CPE
Ergonomist
BJC Health System
St. Louis, Missouri

The authors would also like to thank the safety and health professionals who participated in the focus groups and evaluated the content and format of this document, as well as the healthcare facilities that participated in the hazard demonstration and hazard correction photographs.

Photo Credits

Alex Telfer, www.proofphoto.com, for the use of the nurse image on the cover

Part I. Introduction

Introduction

Slip, Trip, and Fall (STF) Prevention for Healthcare Workers

Work-related slip, trip, and fall incidents can frequently result in serious disabling injuries that impact a healthcare employee's ability to do his or her job, often resulting in

- lost workdays,
- reduced productivity,
- expensive worker compensation claims, and
- diminished ability to care for patients.

According to the U.S. Bureau of Labor Statistics [2009], the incidence rate of lost-workday injuries from slips, trips, and falls (STFs) on the same level in hospitals was 38.2 per 10,000 employees, which was 90% greater than the average rate for all other private industries combined (20.1 per 10,000 employees). STFs as a whole are the second most common cause of lost-workday injuries in hospitals.

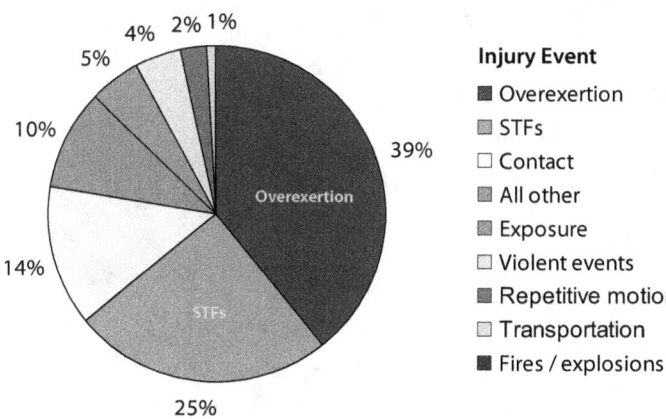

Figure 1. Percent distribution of hospital employee injuries by injury type

An analysis of workers' compensation injury claims from acute-care hospitals [Bell et al. 2008] showed that the lower extremities (knees, ankles, feet) were the body parts most commonly injured after STFs (Table 1) and the nature of injury was most often sprains, strains, dislocations and tears (Table 2). In addition, STFs were significantly more likely to result in fractures and multiple injuries than were other types of injuries.

What is the Purpose of this Workbook?

This workbook identifies the top 10 STF hazards specific to healthcare facilities. For each hazard this workbook will:

1. Explain how the hazard contributes to STFs,
2. Identify where the hazard is likely to occur, and
3. Provide recommendations to reduce or eliminate the hazard.

Slips, trips, and falls are preventable. This workbook provides guidance on implementing a STF prevention program to protect healthcare workers. The goal of the workbook is to familiarize you with common STF hazards in healthcare facilities so you are able to recognize and reduce the risk to employees. Throughout the workbook, pictures show either prevention strategies or hazards in healthcare facilities. Pictures outlined in red are hazards. In addition, both visitors and patients will benefit from an STF prevention program in your facility reducing their risk as well. To further assist you, **a checklist** is provided on **page 35** to help you identify different hazards in your healthcare facility.

Who Should Use the Workbook?

The Workbook is intended for healthcare facility administrators, safety and health professionals, facility managers, housekeeping managers, food service managers, and workers who are responsible for safety.

Table 1. Slip, trip and fall (STF) workers' compensation claims by body part injured, 1996–2005.

Body part	n	% of total STF claims
Lower extremities	185	44.9
Upper extremities	69	16.7
Multiple body parts	67	16.7
Back/trunk	73	16.2
Head/neck	18	4.3
Unknown	60	12.7
Total	472	100.0

Source: Bell et al. 2008

Table 2. STF workers' compensation claims by nature of injury, 1996–2005.

Nature of injury	n	% of total STF claims
Sprains, strains, dislocations, tears	228	48.3
Bruises, contusions, concussions	104	22.0
Fractures	40	8.4
Multiple non-specified injuries	20	4.2
Cuts, lacerations, punctures, abrasions	12	2.5
Burns and scalds (thermal, chemical, electrical)	1	<1
Other injuries	1	<1
Unknown	66	13.9
Total	472	100.0

Source Bell et al. 2008

Research Supports a Slip, Trip, and Fall Prevention Program in Healthcare Facilities

Research conducted by the National Institute for Occupational Safety and Health and its collaborators [Bell et al. 2008] has shown that implementating a comprehensive STF prevention program in hospitals can lead to significant declines in STF-related workers' compensation claims. Researchers worked with hospital staff to design, implement, and evaluate a comprehensive STF prevention program over a 10-year period from 1996 through 2005 in three acute-care hospitals. The hospitals' total STF workers' compensation claims declined 59% after implementating the comprehensive STF prevention program.

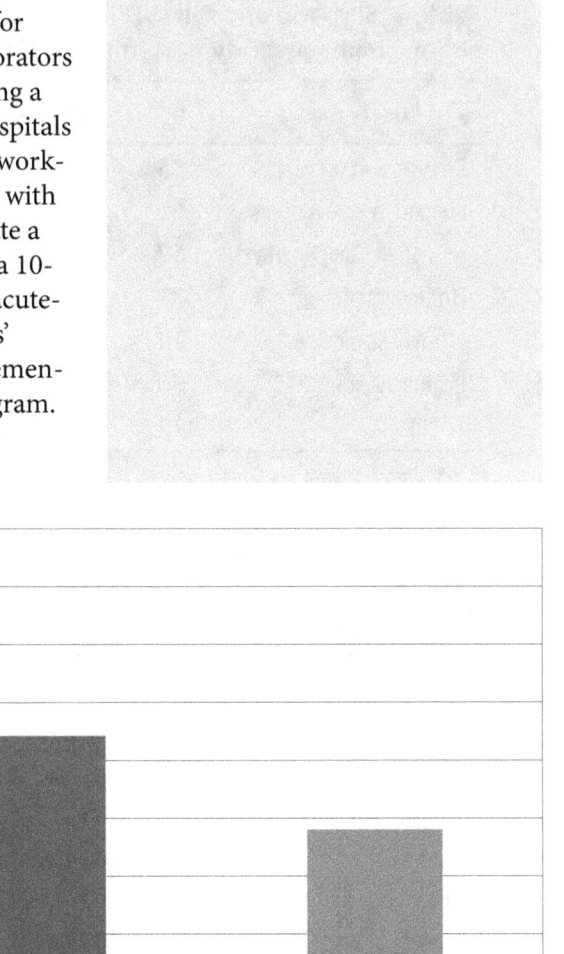

Figure 2. Slip, trip, and fall worker's compensation claim rate by time period.

Part II. Top 10 Hazards

1. Contaminants on the Floor (Water, Grease, Oil, Fluid, Food)

What is the hazard?

Contaminants on the floor are the leading cause of STF incidents in healthcare facilities [Courtney et al. 2006, Bell et al. 2008]. Water, grease, and other fluids can make walking surfaces slippery. Well-documented housekeeping procedures, correct floor cleaning, proper usage of mats and signs, accessible clean-up materials, and slip-resistant shoes will help to minimize the risk of slipping (ANSI 2001).

Figure 1.1. Standing water on floor in dishwashing area

Where does the hazard occur?

- Food services areas: kitchen, cafeteria, serving line, buffet, ice machines, freezers, dishwashers, sinks, and drains (please refer to the drainage section for additional information on page 14) (see Figures 1.1 and 2.3)
- Decontamination area: when wet equipment is transferred from one area to another
- Soap dispensers
- Drinking fountains
- Building entrances, where rain and snow are tracked inside

Figure 1.2. Water on floor caused by pipe splashing water.

Prevention Strategies

Provide and maintain a written housekeeping program.

A written housekeeping program can help ensure the quality and consistency of housekeeping procedures. A copy of the housekeeping program should be provided to all employees and they should know where to find additional copies. **The program should describe**

- How to immediately contact the housekeeping department
- Where and how cleaning materials and products are stored
- When to use wet floor signs and barriers and where signs are stored
- When specific areas of the healthcare facility should be cleaned
- What cleaning methods are appropriate for different areas and surfaces

Keep floors clean and dry.

- Encourage workers to cover, clean, or report spills promptly.
- Hang or place spill pads (see Figures 1.3a, 1.3b, 1.3c), paper towel holders, pop-up-tent wet floor signs (see Figures 1.4a, 1.4b) in convenient locations throughout the healthcare facility so employees have easy access to products to clean, cover, and highlight a spill.
- Advertise phone/pager numbers for housekeeping through emails, posters, and general awareness campaigns.
- Place water-absorbent walk-off mats where water, ice, or soap may drip onto the floor (see Figure 1.5). Use beveled-edge, flat, and continuous mats. (For more information on mats, please refer to the section on *Improper Use of Floor Mats*).
- Provide walk-off mats, paper towel holders, trash cans, and umbrella bags near entrances and water fountains to minimize wet floors (see Figure 1.6, page 12).

Figure 1.3a. Wall-mounted spill pads for use by employees and vistors

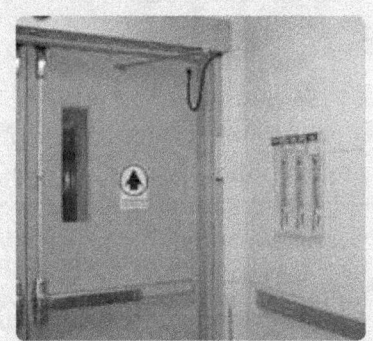

Figure 1.3b. Wall-mounted spill pads

Figure 1.3c. Spill pads for use in building entryway

- Mats should be large enough so that several footsteps will take place on the mat; if there is water around or beyond the mat, it means that the mat is not large enough and/or is saturated and needs to be replaced.
- Secure mats from moving and make sure they have slip-resistant backing. Remind staff to lay mats in the correct position daily, and use visual cues such as tape on the floor if necessary.
- Make sure that drip pans of ice machines and food carts are properly maintained so that water does not spill onto the floor.

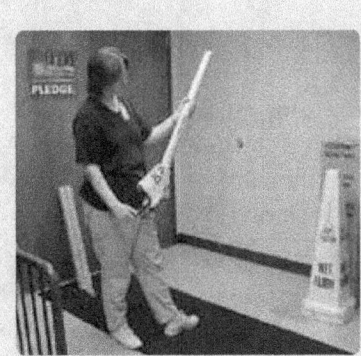

Figure 1.4a. Pop-up tent wet floor sign stored in tube

Use proper cleaning procedures for floors.

- Optimal floor cleaning procedures may prevent slips and falls (Quirion 2004, 2006; Quirion et al. 2008). Research has shown that a two-step mopping process is better than damp-mopping.
- In the two-step process, 1) cleaning solution is applied on a section of the floor with a dripping mop, and 2) after a few minutes, the cleaning solution is removed with a wrung mop, before the solution dries.
- Make sure the cleaning product can be used on common floor contaminants.
- Make sure cleaning products are mixed according to manufacturer's directions.

Figure 1.4b. Pop-up tent wet floor sign stored in tube hung on wall

Wear slip-resistant shoes.

- Choose shoes that are slip-resistant [Loo-Morrey and Houlihan 2007, Collins et al. 2008, Di Pilla 2010]. Slip-resistant shoes are an important component of a comprehensive STF prevention program [Bell et al. 2008]. All healthcare facility employees may benefit from slip-resistant shoes. Food services, housekeeping, and maintenance staff are at greatest risk for a STF due to exposure to water, grease, or slippery walking surfaces.
- Staff that work in areas that are continually wet, such as dishwashing and surgical instrument decontamination areas can benefit from slip-resistant shoes.

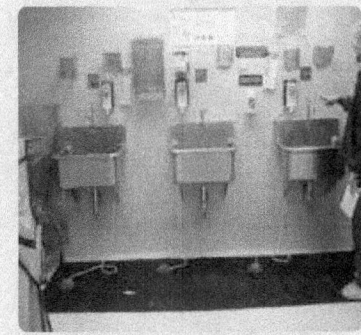

Figure 1.5. Drip-less brush-free cleaning solution provided at sinks minimizes drips and water splashes

- Shoe fit, comfort, and style are important factors that determine whether employees will wear slip-resistant shoes. Employees should have the opportunity to try on shoes to obtain the proper fit before purchasing, or be able to exchange purchased shoes as needed to obtain proper fit.
- Anecdotal evidence suggests that use of slip-resistant shoes by employees is enhanced when the employer shares the cost of approved slip-resistant shoes and provides a payroll deduction option for shoe purchases.

Prevent entry into areas that are wet.

- Use highly-visible caution signs (taller, with flashing lights or signs on top) (see Figure 1.7) to inform employees and visitors to be careful and avoid contaminated area.
- Block off areas during floor cleaning, stripping, and waxing (see Figure 1.8).
- Use barrier products (see Figures 1.9 and 1.10) or caution tape (see Figure 1.11) to prevent employees from entering an area being cleaned or from stepping on a spill.
- Use a long barrier device if a dry lane must be kept clear for passage (see Figures 1.12a and 1.12b).
- Use barrier devices to prevent water and other fluids from entering hallways when cleaning rooms (see Figure 1.13). Use in conjunction with tension bar or other blocking device so the floor barrier does not become a tripping hazard.
- Remove all signs once the floor is clean and dry so they do not become commonplace and ignored by staff.

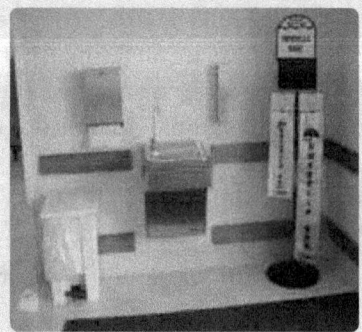

Figure 1.6. Good configuration near building entrance; large absorbent mat, umbrella bags, paper towel dispenser, and trash can

Figure 1.7. High visibility caution sign with warning sign on top or a flashing light on top could be used

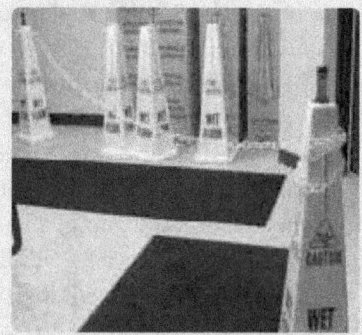

Figure 1.8. Chain to prevent entry between warning signs

Figure 1.9. Chain and warning signs to block access to room during cleaning

Figure 1.10. Spring-loaded tension bar holds warning sign to block access to room during cleaning.

Figure 1.11. Caution wet floor warning tape

Figure 1.12a. Barrier device to block off sections of hallway

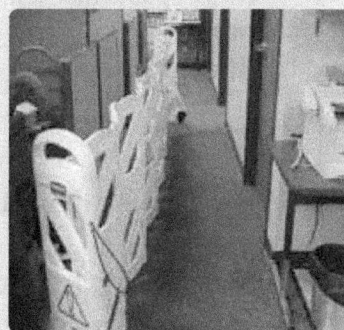

Figure 1.12b. Barrier device to block off sections of hallway

Figure 1.13. Barrier device to block fluids at doorway, use with a tension bar (Figure 1.10) to block access

Slip, Trip, and Fall Prevention 13

2. Poor Drainage: Pipes and Drains

What is the hazard?

Drains and water pipes that are improperly aligned can cause liquid to spill onto walking surfaces, while clogged drains can cause water to back up onto the floor (see Figures 2.1, 2.2, 2.3, 2.4, 2.5a, 2.5b, and 2.5c).

Where does the hazard occur?

- Drains inside the healthcare facility where liquids accumulate (particularly in kitchens and decontamination areas)
- Down spouts that spill rainwater onto sidewalks

Prevention Strategies

- Check that pipes are correctly aligned with the drain they are emptying into.
- Unclog drains regularly, particularly in kitchens.
- Redirect downspouts away from sidewalks with high-pedestrian traffic.

Figure 2.1. Pipe is not aligned properly with the drain causing standing water on the floor

Figure 2.2. Drain is partially clogged with debris

Figure 2.3. Pipe splashing water onto the floor around a drain

Figure 2.4. Drainpipe routing water onto walkway

Figure 2.5b. Drainpipe routing water onto a sloped sidewalk (side view)

Figure 2.5a. Drainpipe routing water onto a sloped sidewalk (front view)

Figure 2.5c. Corrected drainpipe. Water was diverted underground to the sewer pipe, and spout no longer directed onto sidewalk

Slip, Trip, and Fall Prevention   15

3. Indoor Walking Surface Irregularities

What is the hazard?

Damaged, warped, buckled, or uneven flooring surfaces inside healthcare facilities can cause employees to stumble, trip, slip, or fall (see Figures 3.1, 3.2, and 3.3).

Where does the hazard occur?

- Building entrances
- Patient rooms
- Operating rooms
- Hallways
- Around drains in the floor (refer to page 14)
- Floor matting (refer to page 26)

Prevention Strategies

- Replace or re-stretch loose or buckled carpeting.
- Remove, patch underneath, and replace indented or blistered vinyl tile.
- Patch or fill cracks in indoor walkways greater than ¼" wide.
- Reduce or eliminate trip hazards over ¼" high in all areas of pedestrian travel. For changes in level ¼" to ½" high, bevel with a slope no greater than 1:2. For heights greater than ½" high consider a ramp [U.S. Access Board 2003]
- Create visual cues. Highlight changes in walkway elevation with Safety Yellow warning paint [U.S. Access Board 2000].
- Replace smooth flooring materials **in areas normally exposed to water, grease and/or particulate matter** with rougher-surfaced flooring when renovating or replacing healthcare flooring.
- Make sure elevators are leveled properly so that elevator floors line up evenly with hallway floors.

Figure 3.1. Buckled carpet

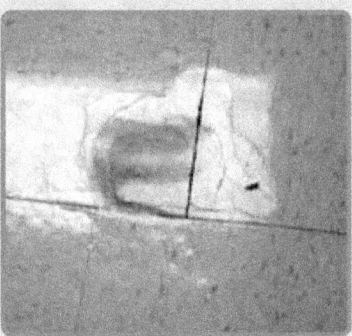

Figure 3.2. Dented vinyl tiles

Figure 3.3. Tile floor where holes were not patched after something was removed

4. Outdoor Walking Surface Irregularities

What is the hazard?

Poorly maintained, uneven ground, protruding structures, holes, rocks, leaves, and other debris can cause employees to stumble, trip, slip, or fall (see Figures 4.1, 4.2, 4.3, 4.4, 4.5, and 4.6, page 18).

Where does the hazard occur?

- Entrances
- Lawns
- Parking garages and lots
- Walkways
- Around drains in the ground (refer to page 14)

Prevention Strategies

- Patch or fill cracks in walkways greater than ½" wide.
- Patch, fill, or repave outdoor areas that have deep grooves, cracks, or holes.
- Create visual cues. Highlight changes in curb or walkway elevation with Safety Yellow warning paint.
- Concrete wheel stops in parking lots can be a tripping hazard and should not be used.
- Remove stones and debris from walking surfaces.
- Ensure that underground watering system structures are covered or highlighted.

Figure 4.1. Holes in grassy area between a parking lot and hospital building

Figure 4.2. Area of sloped pavement that should be highlighted with **Safety Yellow** paint

Figure 4.3. Stones and debris on walking surfaces in a parking lot

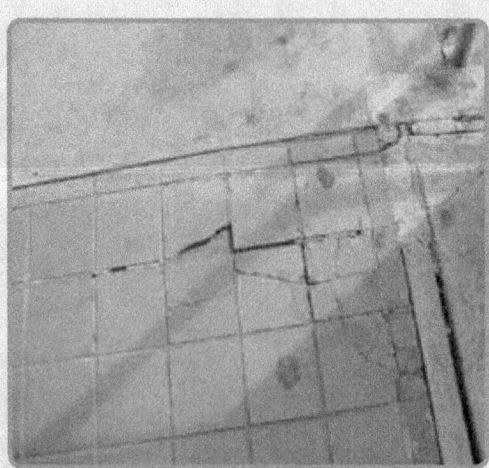

Figure 4.4. Cracked tiles in a walkway outside a hospital building

Figure 4.5. Large separation (greater than ½") in a walkway outside a hospital building

Figure 4.6. Change in elevation (greater than ½") in a walkway outside a hospital building

5. Weather Conditions: Ice and Snow

What is the hazard?
Ice and snow can cause employees to slip and fall.

Where does the hazard occur?
- Entrances
- Parking garages and lots
- Walkways
- Outside stairs

Prevention Strategies
- Have an aggressive program to promptly remove ice and snow from parking lots, garages, and sidewalks.
- Distribute winter weather warnings via email to staff when ice and snow are predicted. For staff that does not have access to email, provide notices on bulletin boards.
- Place freezing weather warning monitors at entrances to employee parking areas (see Figure 5.1).
- Display phone or pager number for maintenance department via posters and emails to encourage employees to report icy conditions.
- Place labeled bins (see Figure 5.2) filled with ice melting chemicals and scoops that anyone can use immediately on icy patches. Consider placing bins in areas of heavy pedestrian traffic such as the top and bottom of outdoor stairways, parking garage exits and entrances, and healthcare facility entrances. The bins should be labeled with the appropriate Material Safety Data Sheets (MSDS) and include instructions for handling ice melting chemicals. Bins should be secured so they cannot be removed.
- Provide additional mats in entrances during winter months and when it rains.
- Consider slip-resistant footwear (including ice cleats) for employees who work or travel outdoors as part of their jobs.

Figure 5.1. Ice Alert temperature monitor

Figure 5.2. Plastic bin to hold ice-melt chemicals and scoop

6. Inadequate Lighting

What is the hazard?
Inadequate lighting impairs vision and one's ability to see hazards. Proper lighting allows employees to see their surroundings and notice unsafe conditions in time to avoid them (see Figure 6.1).

Where does the hazard occur?
- Parking structures
- Storage rooms
- Hallways
- Stairwells
- Walkways both inside and outside the facility

Prevention Strategies
- Install more light fixtures in poorly lit areas.
- Verify light bulbs have an appropriate brightness.
- Install light fixtures that emit light from all sides.

Figure 6.1. Dim lighting in parking area; light fixture is partially obscured by ceiling beams

7. Stairs and Handrails

What is the hazard?

Proper construction and maintenance of stairs and handrails (described in ANSI 2007, NFPA 2002, and below) can reduce hazards. Stairs that are poorly marked or uneven, as well as handrails that are not of the appropriate size, height, or are poorly maintained can lead to missteps and can cause employees to trip and fall (see Figures 7.1, 7.2, and 7.3).

Where does the hazard occur?

- Indoor and outdoor stairs
- Steps inside classrooms or conference rooms
- Elevated and/or sloping walkways
- Parking structures
- Ramps

Prevention Strategies

- Create visual cues. Paint (Safety Yellow or other high contrast paint), tape, or highlight the edge (nosing) of each step, including the top and bottom, to provide a cue of a change in elevation.
- Check that stair treads and nosing are slip resistant and extend the whole tread. This is especially important for outside stairs exposed to the elements or stairways exposed to wet conditions.
- Ensure that stairs are kept free of ice, snow, water, and other slippery contaminants.
- Check that stairwells have adequate lighting.
- Consider adding a handrail at locations that have less than 4 steps (such as employee shuttle bus stop, building entrances, conference theaters).
- Confirm all handrails are within an appropriate height range (34–38" from the stepping surface).

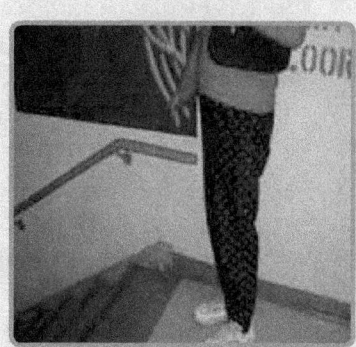

Figure 7.1. Handrail was lowered in order to hang large picture above; handrail too low to be of use

Figure 7.2. Uneven steps by shuttle bus stop, lacking a handrail

Figure 7.3. Discontinuous handrail too low for use by pedestrians

- Check that discontinuous handrails are of a consistent height.
- Check that handrails extend full length of stair and extend 12 inches at top and one tread depth at bottom.
- Check that handrails are available on both sides. For stairs >44 inches wide, two handrails are recommended. For stairs <44 inches than at least one handrail on right side descending stairway. Check local code requirements.
- Check, for open stairways, that a two-rail system is present; a top rail at 42 inches and a second handrail at 34 inches minimum and 38 inches maximum vertically above stair nosings. Protect the open area under the top rail to the stairway steps by installing a fixed barrier.

8. Stepstools and Ladders

What is the hazard?
Stepstools and ladders used to work from heights can create a hazardous situation if not used properly.

Where does the hazard occur?
- Outdoors (refer to page 17)
- Kitchens and pantries
- Pharmacy
- Medical records office
- Areas with elevated storage

Prevention Strategies
- Train employees on the proper use of ladders [OSHA 2003].
- Wear appropriate footwear for climbing; shoes should have a closed back and sufficient tread on the sole to prevent slipping on ladder rungs or steps.
- Place ladders and stepstools on level surfaces before climbing.
- Check that stepladders are fully opened before climbing.
- Maintain three points of contact with the ladder at all time while ascending and descending (two hands and one foot or one hand and two feet).

9. Tripping Hazards: Clutter, Including Loose Cords, Hoses, Wires, Medical Tubing

What is the hazard?

Clutter can build up in storage areas, work areas, hallways, and walkways potentially leading to an STF incident. Exposed cords on the floor, stretched across walkways, and tangled near work spaces can catch an employee's foot and lead to a trip and fall incident (see Figures 9.1, 9.2, 9.3, 9.4, 9.5, 9.6, 9.7, and 9.8, and 9.10a, pages 24–26).

Where does the hazard occur?

- Nursing stations
- Operating rooms
- Patient rooms
- Computer workstations
- Storage areas
- Hallways and walkways
- Work stations

Prevention Strategies

- Organize storage areas to eliminate clutter.
- Consider wall-mounted storage hooks, shelves, hose spools, etc.
- Clear walkways.
- Use cord organizers to bundle cords (see Figure 9.9, page 26).
- Cover cords on floor with a beveled protective cover or tape cords to flooring (see Figure 9.10b, page 26).
- Use retractable cord holders (see Figure 9.11, page 26).
- Mount cords near or underneath the desk.
- Clear walkways and work areas to allow employees to move more freely and safely.

Figure 9.1. Loose phone cord

Figure 9.2. Loose computer cables around an employee workstation and guest chair

Figure 9.3. Unsecured items in a corner; wall-mounted storage could be used for these items

Figure 9.4. Loose medical tubing near an employee's foot

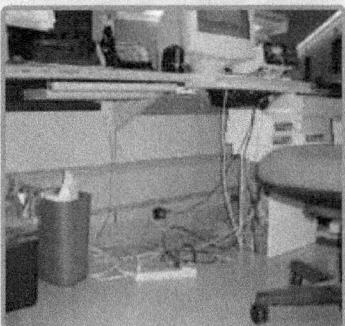

Figure 9.6. Loose cords at nursing workstation

Figure 9.5. Uncoiled hose in a walkway

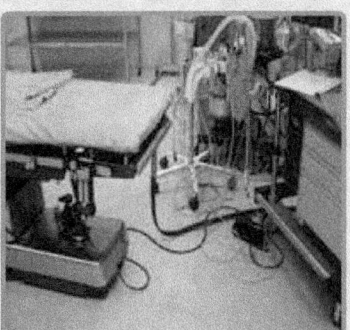

Figure 9.7. Loose cords and cables stretched across pathways in an operating room

Slip, Trip, and Fall Prevention   25

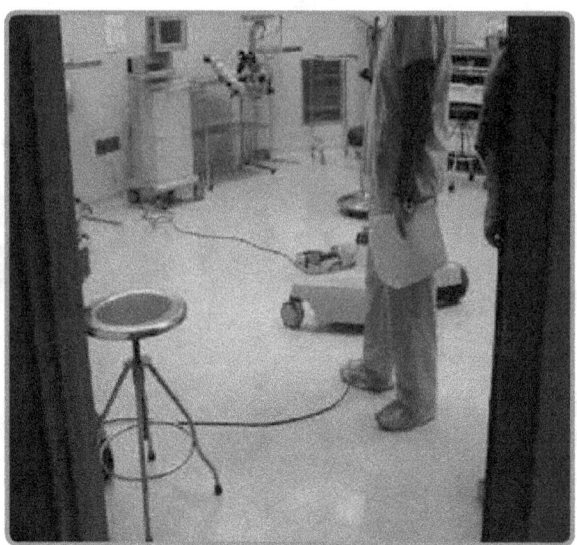

Figure 9.8. Cord stretched across pathway in an operating room (See Brogmus et al. 2007 for suggestions.)

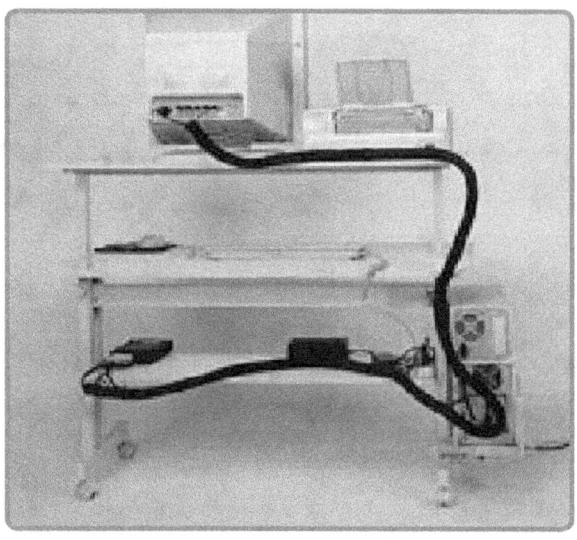

Figure 9.9. Plastic outer sleeve that covers the loose cords (i.e., cord organizer)

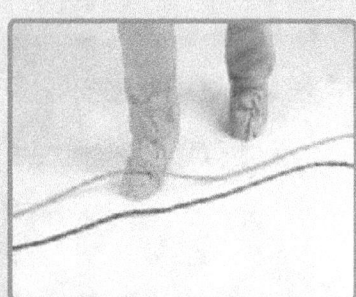

Figure 9.10a. Loose cords on floor

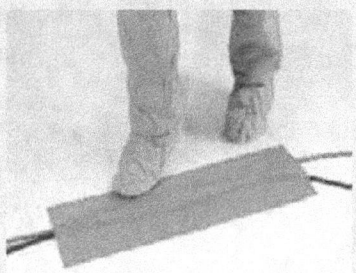

Figure 9.10b. Cords on floor partially secured with cord cover. Cover can extend over the length of the exposed cord.

Figure 9.11. Retractable cord holder

10. Improper Use of Floor Mats and Runners

What is the hazard?

Mats are used to prevent STFs to provide slip-resistant walking surfaces by absorbing liquid, and by removing dirt, debris, and liquid from shoes. **Mats are only effective if properly used and maintained.** Old or poorly placed mats can contribute to slips, trips, and falls (Figures 10.1, 10.2, and 10.3).

Where does the problem occur?

- Healthcare facility entrances
- Food preparation and serving areas
- Under sinks
- Water fountains
- Histology labs

Prevention Strategies

- Mats and runners at healthcare facility entrances should be sufficiently large so that several footsteps fall on the mat, cleaning contaminants off the shoes, before the shoes contact the flooring.
- Place additional mats if necessary in entrances during ice, snow, and rainy conditions. If there is water on the floor beyond the last mat, additional mats or runners may be necessary.
- Use non-slip mats in areas where employees may routinely encounter wet flooring.
- Use beveled-edge, flat, and continuous or interlocking mats.
- Replace mats that are curled, ripped, or worn; secure edges with carpet tape if needed
- Secure mats from moving.
- Paint small markers on the floor to remind staff to lay mats in the correct position everyday.

Figure 10.1. Edge of mat is not flat on the ground

Figure 10.2. Frayed, worn edges on a comfort mat

Figure 10.3. Mat is too small, contaminants present on the floor

Part III. Tools and Resources

Examine Employee Slips, Trips, and Falls Injuries: The First Steps Toward Prevention

Know your healthcare facility's STF history.

Review your healthcare facility's historical injury records for STF incidents. Obtain copies of and check workers' compensation claims, incident reports, first report of employee injury, OSHA and/or occupational health nurse logs. Read the narrative descriptions of the incidents to identify what types of STFs are most common in your healthcare facility and to identify specific locations where multiple STFs or "injury hot spots" may have happened over the years. When a STF incident occurs, carefully examine the circumstances of the incident to see where prevention measures can be implemented.

The following three items may be used to code the specifics of the incident, once it has been identified from the injury narrative as a slip, trip, or fall. These items are adapted from the Bureau of Labor Statistics Occupational Injury and Illness Classification Manual [BLS 2007] and have been tailored for STF events common to healthcare environments. The frequency of the answers to each of these items may be calculated at the end of each year to determine which injury events and hazards are most commonly occurring.

Slips, Trips, or Falls

1. What was the first initiating event?
   - ☐ Slip
   - ☐ Trip (includes caught on)
   - ☐ Loss of balance
   - ☐ Unknown

2. Which choice best describes the STF injury event?
   ☐ A slip or trip that did not result in a fall

   A fall from an elevation, such as
   ☐ A fall while standing on a chair
   ☐ A fall from a ladder or stepstool
   ☐ A fall down stairs or steps
   ☐ A fall from a non-moving vehicle
   ☐ Other fall from an elevation (describe)
   _____

   A same-level fall, such as
   ☐ A fall while walking or working
   ☐ A fall from a chair while sitting
   ☐ A fall while tripping up stairs
   ☐ Other same-level fall (describe)
   _____

   ☐ Unknown

3. Were there any hazards present that may have contributed to the injury event?
   ☐ Contaminant (examples: water, soap, body fluid, grease/oil, coffee, wax, gel, slick, slippery not otherwise classified, etc.)
   ☐ Cord or tubing (examples: hose, medical tubing, phone cord, nurse call cord, equipment cords)
   ☐ Object (examples: objects or items on floor, propped against wall, or in the pathway)
   ☐ Ice or snow
   ☐ Surface irregularity due to buckled, loose, or damaged mat, carpeting, or rug
   ☐ Surface irregularity, other (examples: some part of the walking surface is irregular, cracked tiles, loose gravel, leaves, door guard, drain dip, utility hole in floor, hole in lawn)
   ☐ A curb or wheel stop
   ☐ Bodily reaction (examples: awkward posture,

reaching, crouching, bending, carrying something, patient or object handling, or just stated as they "fell")
- ☐ Lack of space/restricted pathway
- ☐ Steps, stairs, or handrail
- ☐ Chair or stool
- ☐ Lighting
- ☐ Inappropriate or malfunctioning footwear
- ☐ Unknown / no specific hazard mentioned
- ☐ Other (please specify) _____

_____

Check for STF hazards on a regular basis.

Conduct regular walk throughs using the slips, trips, and falls **checklist** provided on page 35 of this Workbook to identify and record hazards to be addressed. Hazards should be photographed, described, and kept on file so that changes can be made and documented. In the comments section, it is helpful to list a person responsible for fixing the hazard and a targeted completion date.

Employee Communication: Training and Involvement

All healthcare facility employees are at risk, therefore all employees should be trained on how to recognize STF hazards, and be involved in the development and implementation of prevention strategies.

It is important to have written housekeeping procedures that require all employees (including direct patient care staff such as nurses) to immediately report spills/snow/ice etc. to initiate a prompt response by the housekeeping or facilities departments.

- Make cleaning and safety supplies and products easily accessible to all staff.
- Incorporate slip, trip, and fall awareness and prevention into routine safety training.
- Conduct General Awareness campaigns within the healthcare facility (i.e., booths, posters, emails, paycheck inserts, and incentives) educating employees about the risk of STFs at work and what they can do to prevent injuries.
    — Consider making key chains or something employees can carry with them that have emergency numbers for housekeep to quickly report floor contaminations or hazards.
- Reinforce the use of prevention equipment (handrails and appropriate footwear for example) frequently with staff.
- **Track Success:** Provide feedback to employees on how the facility is doing with regard to STF injury rates.

Slips, Trips, and Falls Checklist

Read each statement and place a check mark in the box indicating either Yes or No. If a check mark falls in a red-shaded box, that indicates a hazardous condition may be present and needs further attention. Refer back to Section II for the top 10 hazards and prevention strategies.

Hospital Slip, Trip, and Fall Hazard Checklist				
Contamination and Irregularities (Indoor Walking and Working Surfaces)	Yes	No	Locations / Comments	Who is Responsible?
Do tiles, linoleum, or other flooring have holes, cracks, or bumps?				
Is carpeting buckled, loose, or frayed?				
Are carpet edges curled up?				
Does floor feel greasy or slippery?				
Are liquid contaminants present (water, grease, oil, cleaning solutions, coffee, body fluids)?				
Are dry contaminants present (powder, sawdust, dirt, flour, food, wax chips)?				
Are there sudden changes in indoor floor elevation > 1/4"?				
Are there metal grates or mesh flooring in the walkway?				
Are water absorbent walk-off mats used in entrances?				
Are slip-resistant mats used in wet areas?				

Contamination and Irregularities (Outdoor Walking and Working Surfaces)	Yes	No	Locations / Comments	Who is Responsible
Are there gaps, cracks, or holes in the outdoor walkway > 1/2"?				
Are there metal grates or mesh flooring in the walkway?				
Is the walkway uneven, with abrupt changes in level > 1/2"?				
Is there debris (pebbles, rocks, leaves, grass clippings) on the walkway?				
Are there any slippery conditions present (water, grease, ice, snow)?				
Are concrete wheel stops in the parking areas highlighted with paint?				

Drainage: Pipes and Drains	Yes	No	Locations / Comments	Who is Responsible?
Are drains clogged or filled with debris?				
Are pipes splashing water onto a walking surface?				
Are outside drain pipes or down spouts spilling water on walkways?				
Are pipes properly aligned with drains inside and outside of the hospital?				

Weather Conditions (Ice and Snow)	Yes	No	Locations / Comments	Who is Responsible?
Are bins containing ice melting chemicals and scoops provided at areas of heavy pedestrian traffic?				
Are ice-melting chemicals swept up once walkways are dry?				
Is winter weather communications distribution system in place?				
Is snow removal appropriately scheduled?				

Stairs and Handrails	Yes	No	Locations / Comments	Who is Responsible?
Are all handrails 34–38" from the floor?				
Are handrails provided on slopes, ramps, stairs?				
Do handrails extend at least as far as the last step?				
Are handrails provided at steps (employee shuttle bus stop, entrances, conference and training rooms)?				
Are the edges /noses of each step painted or marked?				
Are stairway risers and steps all of uniform size?				

Tripping Hazards (Clutter, Loose Cords, Hoses, Wires, and Medical Tubing)	Yes	No	Locations / Comments	Who is Responsible?
Are cords bundled using a cord organizer?				
Are cords on the floor covered with a beveled protective cover or tape?				
Are cords mounted under the desk or on equipment?				
Are hallways, stairs, and walkways clear of clutter (boxes, cords, equipment)?				
Is there appropriate storage (closet, shelves, hooks, lockers)?				
Are stepstools available for use in areas with overhead storage?				
Do rolling office chairs have a sturdy base (no less than 5 legs)?				

Lighting (Check both inside and outside the healthcare facility.)	Yes	No	Locations / Comments	Who is Responsible?
Are light bulbs burned out?				
Are any areas dim, poorly lit, or shadowy?				
Are lighting levels compliant with local codes, ANSI, and/or Illuminating Engineering Society (IESNA) recommendations?				

Mats	Yes	No	Locations / Comments	Who is Responsible?
Do mats have abrupt squared-off edges, lacking a bevel?				
Are mat edges curled up or flipped over?				
Do mats slide around on the floor?				

Slip-Resistant Shoes	Yes	No	Locations / Comments	Who is Responsible?
Are employees wearing slip-resistant shoes (safety shoe marketed as slip-resistant)?				
Do shoe soles have worn-down tread?				
Is the shoe sole tread clogged with dirt, food, debris, or snow?				
Are employees that must work outside wearing slip-resistant footwear?				

Visual Cues	Yes	No	Locations / Comments	Who is Responsible?
Are changes in walkway elevation highlighted?				
Are curbs highlighted?				
Are highly visible wet floor signs available and used correctly?				
Are barriers available and used to prevent access into wet or dangerous areas?				
Are wet floor signs removed promptly once floor is dry / clean?				

Safety Products	Yes	No	Locations / Comments	Who is Responsible?
Are the following products available and conveniently located throughout the hospital?				
Wall-mounted spill absorbant pads or paper towels?				
Cups near water fountains?				
Trash cans?				
Pop-up tent floor signs?				
Umbrella bags?				
Barrier and access restriction devices?				

Employee Communication (Training and Employee Involvement)	Yes	No	Locations / Comments	Who is Responsible?
Do all employees know the contact number for the housekeeping department?				
Are winter weather warnings distributed to staff through email?				
Are all employees aware of the housekeeping procedures?				
Do employees know where safety products are stored?				
Are cleaning methods for all floors and paths recorded and displayed?				
Are employees that use ladders trained in safe ladder use and maintenance?				

Literature Cited

ANSI [2006]. American national standard: Standard for the provision of slip resistance on walking/working surfaces. New York: American National Standards Institute. ANSI/ASSE A1264.2-20061 [www.ansi.org].

ANSI [2007]. Safety requirements for workplace walking working surfaces and their access: floor, wall and roof openings, stairs and guardrail systems. American National Standards Institute. ANSI/ASSE A1264.1-2007 [www.ansi.org].

Bell JL, Collins JW, Wolf L, Grönqvist R, Chiou S, Chang W-R, Sorock GS, Courtney TK, Lombardi DA, Evanoff B [2008]. Evaluation of a comprehensive slip, trip, and fall prevention programme for hospital employees. Ergonomics 51(12):1906–1925 [http://www2a.cdc.gov/nioshtic-2/BuildQyr.asp?s1=20034734&f1=NN&Startyear=&terms=3&Adv=1&ct=&B1=Search&Limit=500&Sort=DP+DESC&D1=10&EndYear=&PageNo=1&RecNo=1&View=f].

Brogmus G, Leone W, Butler L, Hernandez E [2007]. Best practices in OR suite layout and equipment choices to reduce slips, trips, and falls. AORN (Association of PeriOperative Registered Nurses) J 86:384–398.

BLS [2007]. Occupational Injury and Illness Classification Manual (OIICS). Washington, DC: U.S. Department of Labor, Bureau of Labor Statistics.

BLS [2009]. Incidence rates for nonfatal occupational injuries and illnesses involving days away from work per 10,000 full-time workers by industry and selected events or exposures leading to injury or illness, 2008. Washington, DC: U.S. Department of Labor, Bureau of Labor Statistics [http://www.bls.gov/iif/oshcdnew.htm].

Courtney TK, Lombardi DA, Sorock GS, Wellman HM, Verma S, Brennan MJ, Collins J, Bell J, Chang WR, Gronqvist R, Wolf L, DeMaster E, Matz M [2006]. Slips, trips and falls in U.S. hospital workers—detailed investigation. World Congress of the International Ergonomics Association (IEA), Maastricht, Netherlands.

Collins JW, Bell JL, Gronqvist RA, Courtney TK, Lombardi DA, Sorock GS, Chang W-R, Wolf L, Chiou S, Evanoff B, Wellman H, Matz M, Nelson A [2008]. Multi-disciplinary research to prevent slip, trip, and fall (STF) incidents among hospital workers. Contemporary Ergonomics 2008, pp. 693–698.

DiPilla S [2010]. Slip and fall prevention: a practical handbook. 2nd Edition. CRC Press, Boca Raton, FL.

Gielo-Perczak K, Maynard WS, DiDomenico A [2006]. Multidimensional aspects of slips, trips, and falls. In: ed. Robert Williges, ed. Reviews of Human Factors and Ergonomics, HFES. Vol. 2, Santa Monica, CA, pp. 165–194.

Loo-Morrey M, Houlihan R [2007]. Further slip-resistance testing of footwear for use at work. British Health and Safety Executive, Health and Safety Laboratory, Report Number HSL/2007/33. Harpur Hill, Buxton, Derbyshire. SK17 9JN [http://www.hse.gov.uk/slips/research/footwear.htm].

NFPA [2002]. NFPA 101B: Code for means of egress for buildings and structures. National Fire Protection Agency [www.nfpa.org].

OSHA [2003]. Stairways and ladders: a guide to OSHA rules. Washington, DC: U.S. Department of Labor, Occupational Safety and Health Administration. OSHA 3124-12R [http://www.osha.gov/Publications/osha3124.pdf].

Quirion F [2004]. Floor cleaning as a preventive measure against slip and fall accidents. Institut de recherché Robert Sauvé en santé et en sécurité du travail (IRSST) Technical Guide RF-366.

Quirion F [2006]. Beware! slippery floors. Institut de recherche Robert Sauvéen santé et en sécurité du travail (IRSST) [http://www.qinc.ca/entretien/index1.html].

Quirion F, Poirier P, Lehane P [2008]. Improving the cleaning procedure to make kitchen floors less slippery. Ergonomics *51*(12):2013–2029 [http://www.informaworld.com/smpp/942440375-54392648/content~db=all~content=a904367820].

Sotter G [2000]. Stop slip and fall accidents. Sotter Engineering Company, Mission Viejo, CA.

U.S. Access Board [2000]. Detectable warnings: synthesis of U.S. and international practice. Washington, DC: United States Access Board [http://www.access-board.gov/research/DWSynthesis/report.htm].

U.S. Access Board [2003]. Technical bulletin: ground and floor surfaces. Washington, DC: United States Access Board [http://www.access-board.gov/adaag/about/bulletins/surfaces.htm].

Notes

Notes

Notes

www.ingramcontent.com/pod-product-compliance
Lightning Source LLC
Chambersburg PA
CBHW081905170526
45167CB00007B/3163